NEXTGEN LEARNERS PRESENTS

Think Like a Brain!™ Collection

Olly Occipital & The Lookout Tower

How Your Brain Helps You See the World

BOOK 4

Think Like a Brain!™ Collection Overview

Think Like a Brain!™ is a wildly engaging, silly-smart children's book collection that invites kids to explore how their brains work from the inside out. Through playful characters, imaginative settings, and rhythm-rich storytelling, this collection teaches neuroscience, social-emotional learning (SEL), and self-understanding in ways children absorb without even realizing they are learning.

Instead of telling children how to behave or what to feel, Think Like a Brain!™ shows them why their brains respond the way they do. Each story transforms complex brain science into friendly characters and relatable moments, helping kids feel capable, curious, and empowered rather than confused or ashamed.

Designed for children first and embraced by parents and educators as preventative education, this collection supports:

- Emotional regulation and self-awareness
- Executive functioning skills
- Neurodivergent understanding and inclusion
- Brain-body connection and growth
- Compassion for self and others

With humor, heart, and factual accuracy at its core, Think Like a Brain!™ helps children build the skills many adults wish they had learned sooner, one playful page at a time.

One of the collection's most powerful features is how the characters become real-life reminders for children long after the story ends. Once kids learn who the characters are and what jobs they do, they can begin to recognize those same processes in themselves. A big feeling might become, "It looks like Amy is trying to tell you something," or a moment of pause might sound like, "What does your Fred think?" By naming experiences through familiar characters, children learn to notice what is happening inside their brains, reflect on it, and move forward more quickly with understanding rather than frustration.

Series 1:
The Brainy Bunch (Brain Anatomy)

Series Overview: Meet the brain's core parts as a friendly team. Kids learn what each part does and how they work together to help thinking, feeling, moving, and learning.

1. **The Brainy Bunch**
   Focus: Brain overview and teamwork
2. **Brainstem Benny & The Automatic Adventure**
   Focus: Breathing, heartbeat, survival basics
3. **Bella Cerebellum's Balance Ballet**
   Focus: Balance, coordination, practice
4. **Olly Occipital & The Lookout Tower**
   Focus: Vision, visual processing
5. **Tammy Temporal & The Soundy Sound Storm**
   Focus: Hearing, language, sound processing
6. **Perry the Parietal & The Body Map Bash**
   Focus: Touch, space, body awareness
7. **Hippocampus Hal & The Memory Maze**
   Focus: Memory, learning, recall
8. **Amy the Amygdala & The Overactive Alarm**
   Focus: Emotional alarm system
9. **Luna Lefty & Roger Righty's Mixed Up Masterpiece**
   Focus: Left/right hemispheres, teamwork
10. **Callie the Connector & The Brain Bridge**
    Focus: Corpus callosum, brain-to-brain communication

Series 2:

Cortex Castle (Frontal Lobe Executive Functions)

Series Overview: A lively castle of executive-function helpers. Kids practice planning, starting, stopping, focusing, flexible thinking, time skills, and self-talk, with Frontal Fred guiding thoughtful choices.

1. **Cortex Castle**
   Focus: Executive function overview
2. **Frontal Fred & The Decision Dilemma**
   Focus: Decision-making, prioritizing
3. **Impulsive Ian & The Button Box**
   Focus: Impulse control, pause
4. **Memory Molly & The Wandering Thoughts**
   Focus: Working memory, holding steps
5. **Focus Frannie & The Great Distraction Chase**
   Focus: Attention, refocusing
6. **Planella & The Checklist Parade**
   Focus: Planning, sequencing
7. **Flexi & The Changey-Changeroo Day**
   Focus: Cognitive flexibility
8. **Self-Talk Sally & The Thoughtful Echoes**
   Focus: Self-talk, emotional regulation
9. **Starter Stacy & Finisher Finn**
   Focus: Task initiation, task completion
10. M**ilo the Minute Keeper & Mae the Moment Seeker**
    Focus: Time awareness, pacing
11. **The Power of Pausing with Penny**
    Focus: Pause skills, regulation

Series 3:
Busy Brain Café (Neurodivergent Affirmations)

Series Overview: A welcoming café where different brain styles are celebrated. These stories affirm neurodivergent experiences while teaching self-understanding, communication, and compassion for others.

1. **Nathan Neuron & The Bouncy Brain**
   Focus: ADHD metaphor, movement, attention
2. **Sara Sprinkle the Sensa & The Super Sensitive Day**
   Focus: Sensory sensitivity, overload tools
3. **Overwhelmed Owen & The Senses Overload**
   Focus: Sensory overload, coping strategies
4. **Interoceptive Ivy & The Inside Signals**
   Focus: Interoception, body signals
5. **Zane's Zigzag Thoughts & The Path to Peace**
   Focus: Anxiety, grounding
6. **The Busy Brain Café**
   Focus: Neurodivergent inclusion, different processing styles
7. **The Listening Light**
   Focus: Nonverbal communication, understanding
8. **The Language of Love**
   Focus: Connection, communication, compassion
9. **Trying Theo & The How-To Hiccup**
   Focus: Persistence, frustration tolerance
10. **Drowsy Drew & The Nighttime Repair**
    Focus: Sleep, brain repair
11. **Burned-Out Bonnie & The Rebuild**
    Focus: Burnout, rest, recovery

Series 4:
Steady Systems Squad (Regulation & Growth)

Series Overview: A squad that explains how brains regulate, reset, rest, and grow stronger over time, including stress waves, sleep crews, sensory filtering, and rebuilding after hard days.

1. **The Steady Systems Squad**
   Focus: Regulation systems overview
2. **Timothy Thalamus & The Traffic Tower**
   Focus: Sensory filtering, attention routing
3. **Danny Dimmer & Wake-Up Wonda**
   Focus: Arousal and alertness
4. **The Start, Stop, and Switch Station**
   Focus: Initiation, inhibition, shifting
5. **The Night Shift & The Sleepy Crew**
   Focus: Sleep systems, restoration
6. **The Stress Wave & The Return to Calm**
   Focus: Stress response, recovery
7. **The Brain Growth Gathering**
   Focus: Brain growth and development
8. **Paul's Practice Path**
   Focus: Neuroplasticity
9. **Daisy's Daydream Network**
   Focus: Daydream network, imagination
10. **Remy's Rebuild Crew**
    Focus: Recovery and healing

NEXTGEN LEARNERS PRESENTS

Think Like a Brain!™ Collection

Olly Occipital & The Lookout Tower

How Your Brain Helps You See the World

BOOK 4

WRITTEN & ILLUSTRATED BY:

MAISEL MCLAULA

NEXTGEN LEARNERS PRESS™
CINCINNATI, OHIO

This book is published by NextGen Learners Press, an imprint of NextGen Learners LLC.

Maisel McLaula is a pen name used by the author of this work.

First Edition: 2026
Printed in the United States of America
ISBN: 978-1-972860-03-8

This book is intended for educational and informational purposes only. It is not intended to replace professional medical, psychological, or therapeutic advice, diagnosis, or treatment.
Always seek the advice of a qualified professional with any questions regarding a medical or mental health condition.

Artwork and visual elements in this book were created using a combination of original artwork, with enhanced original direction, digital illustration tools, and AI-assisted creative technologies.

A Note for My Readers:

Hi friend,
Your brain is incredible. It thinks, feels, remembers, solves, dances, creates, and so much more!
This series is here to help you understand how your brain works, and to cheer you on as you grow your own thinking superpowers.

Every part of your brain has a job.
Sometimes they get mixed up. That's okay!

When you learn what each part does, you can work with your brain instead of feeling like it's working against you.
So read, giggle, learn, and maybe even draw your own Brainy Bunch when you're done.

I'm cheering you on with every rhyme.

-Maisel McLaula

THINK LIKE A BRAIN!™

BRAIN ANATOMY COLLECTION

EXPLORE THE CASTLE INSIDE YOUR HEAD! MEET EACH ROOM AND ITS BRILLIANT RESIDENT, FROM FRONTAL FRED TO BENNY THE BRAINSTEM. WHIMSY, WONDER, AND NEURONS AWAIT!

For Khaleesi,
Who taught me that seeing differently can become a superpower. I am proud of you for learning to see in your own way and noticing the magic around you everywhere.

And...

To the beautifully wired, the curiously wired, and the wonderfully neurodivergent minds, may you always see your own sparkle clearly.

You are exactly what a great brain looks like.

-Maisel McLaula

In the Lookout Tower, way up in the back,
Lived Olly Occipital, on a visual track.
She watched all the world through a lens made of light,
And helped turn what you saw into clear mental sight.

She noticed the colors, the shadows, the shapes,
The squiggles on papers, the hills and the lakes.

She'd whisper, "That's square!" or "Look! It's a frog!"
And point out the pebble or squirrel or log.

But sometimes too much came into her nest,
The blinking, the flashing, the sparkle obsessed.

When the room got too loud with visual sound,
Olly would gasp and start spinning around.

Her eyes jumped around as the pictures all raced,
She couldn't tell what belonged in what place.

"Too much to look at!" she'd wobble and shout,
"I need a clear path to help figure it out!"

Fred flew up fast with his flashlight and guide,
And offered Olly a place to reside.

"Let's give you some tools to sort what you see,
And make vision work more peacefully."

Fred taught her to slow and to look just one way,
To choose a small target and let the rest fade.

"Pick one thing to focus on, hold it in sight,
Then gently ignore what's too busy or bright."

Olly sighed with relief as her eyes took a rest,
She saw shapes and signs without all the stress.

She could now scan a room and find just one thing,
Like a crayon or sock or a missing earring.

She practiced with puzzles and patterns and light,
And trained her brain to match left from right.

She followed the shapes with her sharp, careful eye,
And drew what she saw as the world wandered by.

Now Olly still sees and observes all day long,
But she filters with care and feels steady and strong.

She doesn't get dizzy or frazzled or stuck,
She's learned that focus takes practice and luck.

She guides the Brainy Bunch when they need to explore,
She points out the clues on each map and each door.

They call on her vision to help find the way,
Whether nightfall or noon or the middle of day.

Olly Occipital now sees without fear,
She's still got her sparkle but also some gear.

From filters to maps and a pause when unsure,
She sees with precision and focus secure.

So when your own vision gets fuzzy or loud,
Or you're lost in the middle of a sensory crowd,

Remember Olly and blink a few times,
Then look with intention on things that make you feel fine.

Olly's Vision Tools

- Use calm colors and limit bright light
-
- Take visual breaks when things feel too tight
-
- Practice tracking with puzzles or games
-
- Try to label the shapes and give them names

What Is the Occipital Lobe?

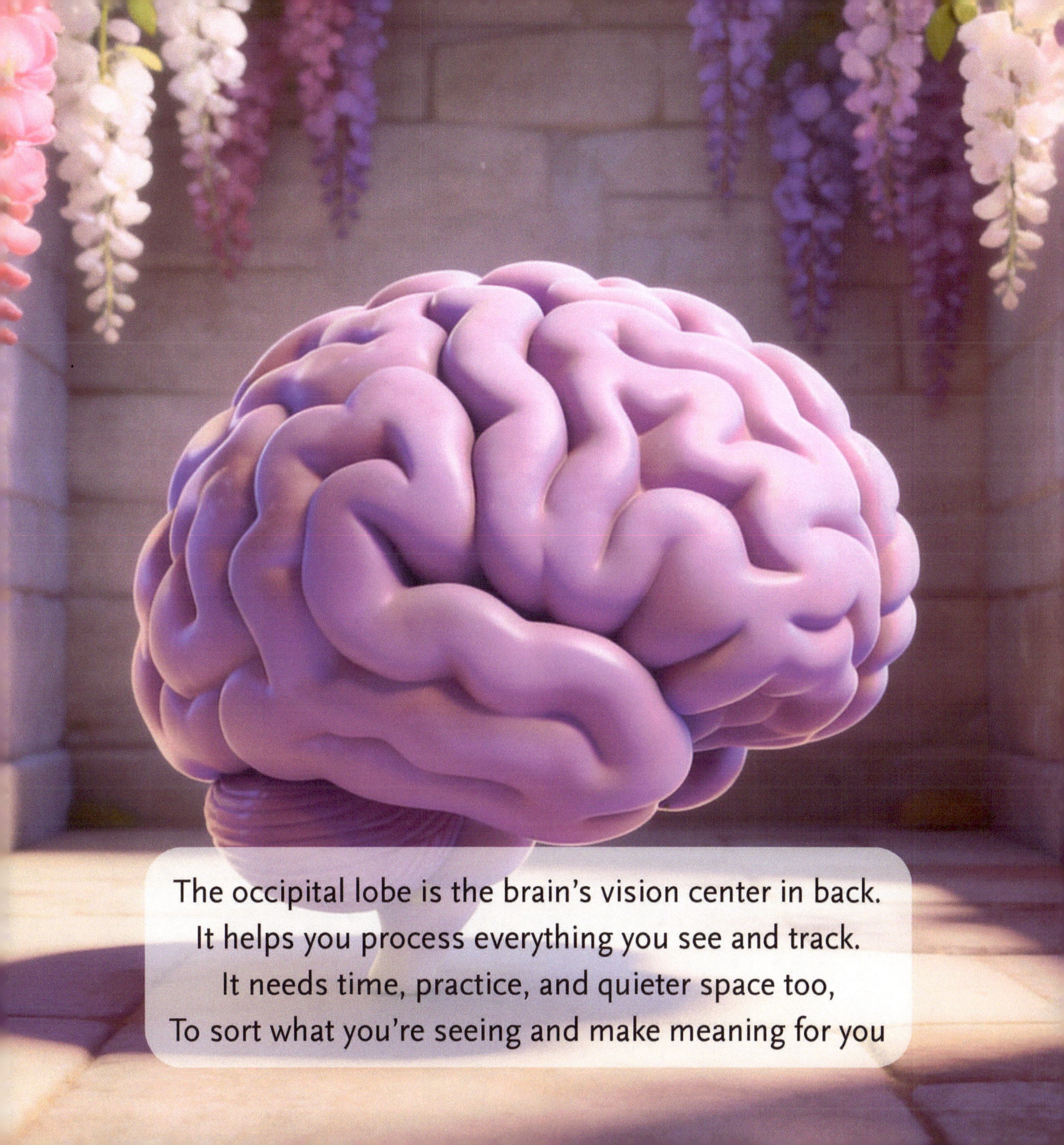

The occipital lobe is the brain's vision center in back.
It helps you process everything you see and track.
It needs time, practice, and quieter space too,
To sort what you're seeing and make meaning for you

Try It Yourself!

- Draw a View Map like Olly to see
- Circle the colors and shapes you find with glee
- Trace a path from desk to door
- Blink three times and notice what's on the floor

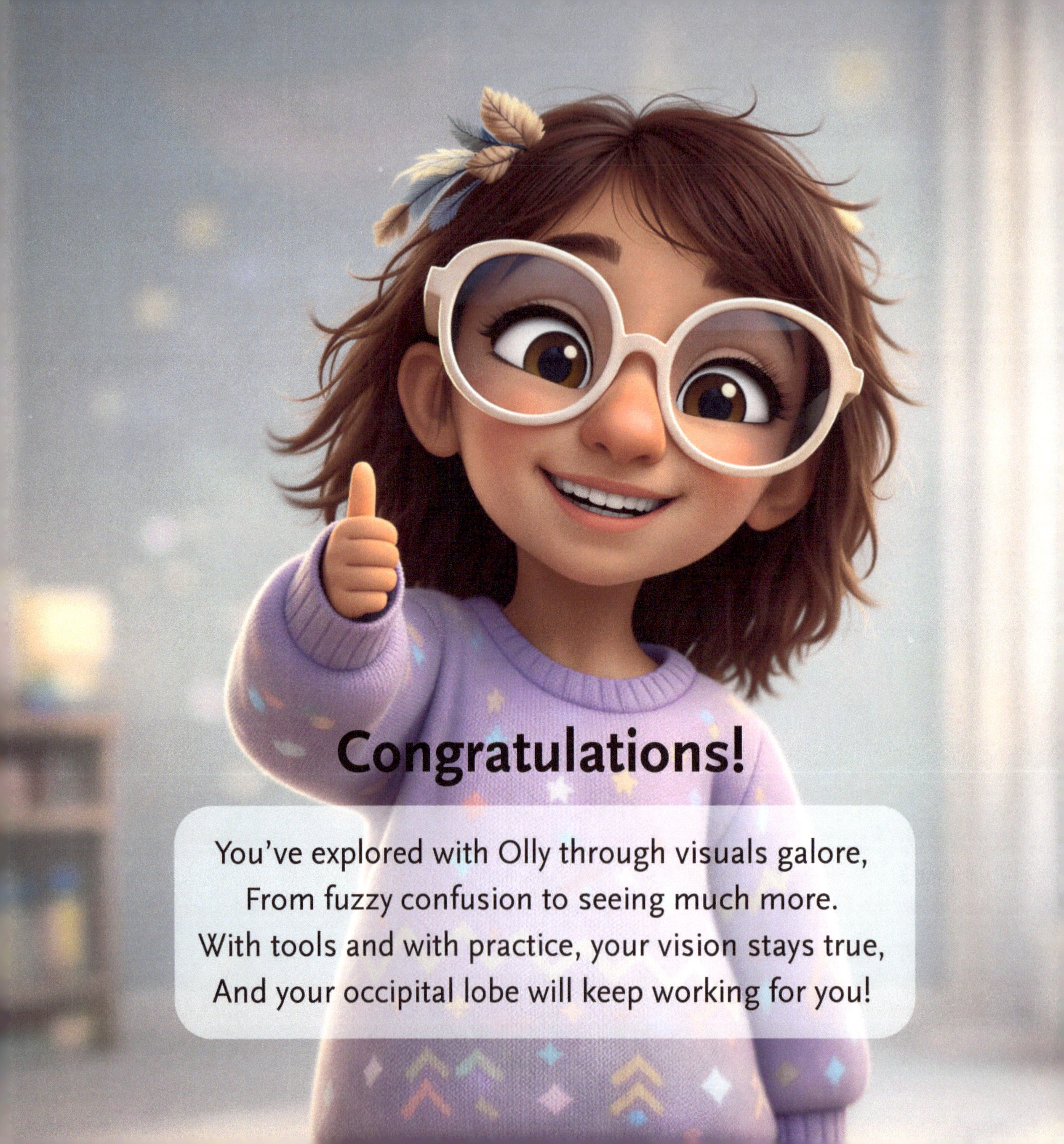

Congratulations!

You've explored with Olly through visuals galore,
From fuzzy confusion to seeing much more.
With tools and with practice, your vision stays true,
And your occipital lobe will keep working for you!

Thank You to My Readers

Thank you for reading this book.
Whether you're a parent, a teacher,
a therapist, or a brilliant young
brain explorer,
your interest in understanding
how the mind works
means the world.

Sharing these books
means spreading compassion and
curiosity.

You are the reason
these stories come alive.
Thank you for believing in the
power of playful learning.

Stay curious. Stay kind.
Stay brainy!

Olly Occipital's *Teacher & Parent Guide*

💡 What This Story Teaches:

- Brain part: Occipital Lobe (vision and visual processing)

- Core idea: Seeing clearly takes focus, filtering, and rest.

- Skills supported: visual attention, tracking, visual discrimination, visual calm

- Big message: Too much visual input can overwhelm the brain. Learning to focus helps everything feel clearer.

Olly shows children that vision is not just about eyes. It is about how the brain organizes what we see.

🧠 Key Concepts for Grown-Ups

Olly represents the occipital lobe, located at the back of the brain. This region processes visual information such as color, shape, movement, contrast, and spatial relationships.

When environments are visually busy (bright lights, clutter, screens, movement everywhere), the occipital system can become overloaded. This may look like:

- Difficulty focusing
- Avoiding eye contact
- Losing place while reading
- Visual fatigue
- Irritability in bright or crowded spaces
- Trouble finding objects

Important foundations:

- Visual processing requires filtering, not just looking.
- Busy visuals can dysregulate the nervous system.
- Children may need visual breaks just like movement breaks.
- Calm visual input supports attention, learning, and emotional regulation.

This story teaches children that:

- It's okay to feel overwhelmed by what you see.
- Focusing on one thing at a time helps.
- Resting your eyes is a healthy brain skill.

🗣 Discussion Starters

- What made Olly feel overwhelmed?
- What helped Olly see more clearly?
- What kinds of places feel too busy for your eyes?
- What helps your eyes feel calm?
- Can you practice looking at just one thing right now?

Encourage children to describe visual experiences without labeling them as good or bad.

👪 Try This Together:

1. One-Thing Focus

Place one object on the table. Invite children to look only at that item for 10-20 seconds. Notice color, shape, and details.

2. Visual Scan

Slowly scan the room and name three things you see. Then close eyes briefly and rest.

3. Tracking Play

Hold a small object and move it slowly from side to side. Have children follow it with their eyes while keeping their head still.

4. Calm Corners

Create a low-clutter space with soft colors and minimal visual input.

5. Visual Breaks

Every 20-30 minutes, look away from screens or work and focus on something far away.

🎯 Bonus Challenge

Build Olly's Lookout Tower

Invite children to draw or build a tower using blocks or paper.

Ask:

- What does Olly look for from her tower?
- What helps her see clearly?
- What should be inside the tower to help eyes rest?

Optional Labels:

- Focus Window
- Blink Button
- Calm Light
- Quiet View

Playful Extension:

I Spy with Olly

Play I Spy using only one category at a time (colors, shapes, or sizes). This teaches selective attention instead of visual overwhelm.

Take turns being Olly. Encourage children to recognize a variety of objects, both easy and hard to spot!

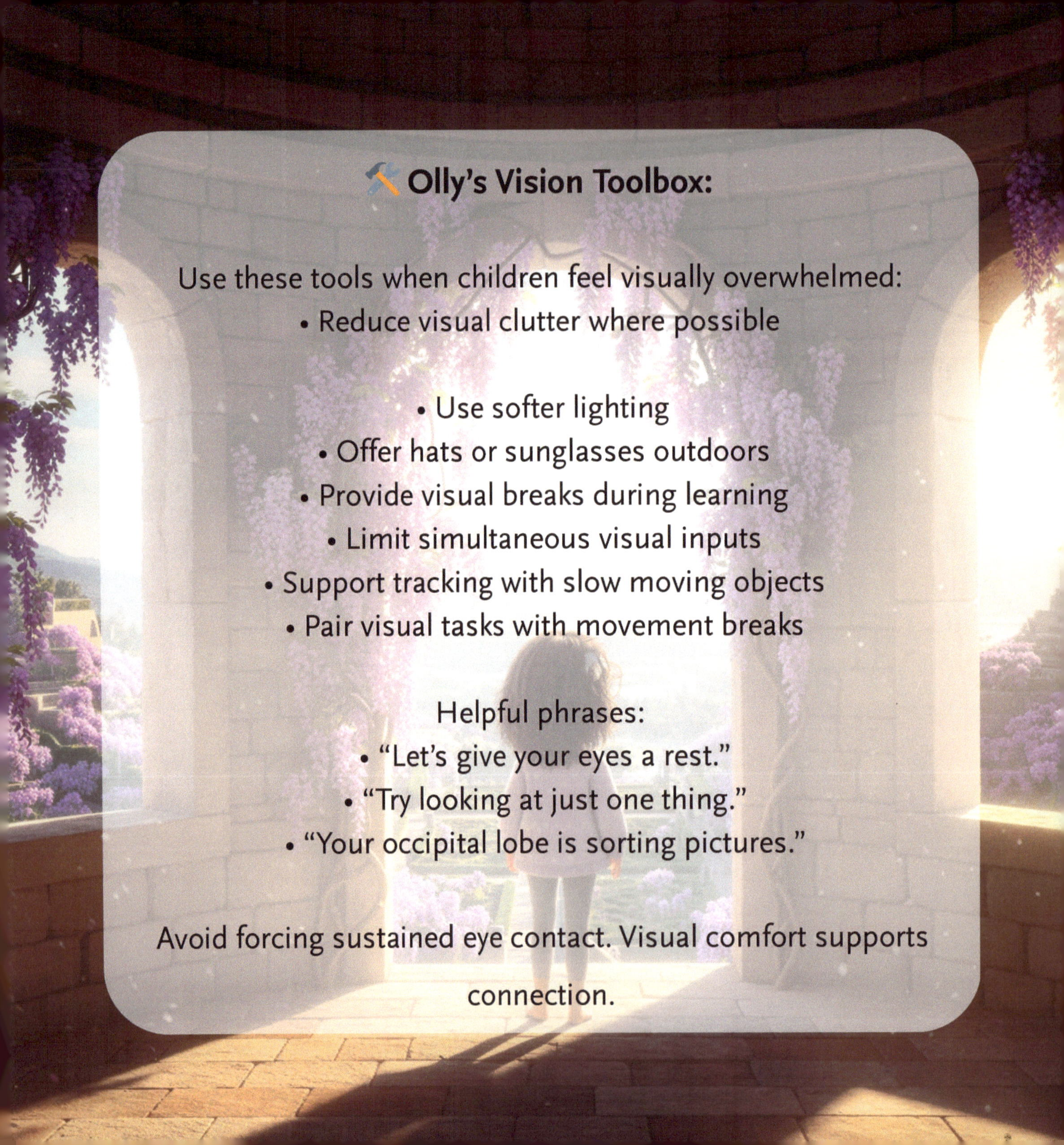

⚒️ Olly's Vision Toolbox:

Use these tools when children feel visually overwhelmed:

- Reduce visual clutter where possible

- Use softer lighting
- Offer hats or sunglasses outdoors
- Provide visual breaks during learning
- Limit simultaneous visual inputs
- Support tracking with slow moving objects
- Pair visual tasks with movement breaks

Helpful phrases:

- "Let's give your eyes a rest."
- "Try looking at just one thing."
- "Your occipital lobe is sorting pictures."

Avoid forcing sustained eye contact. Visual comfort supports connection.

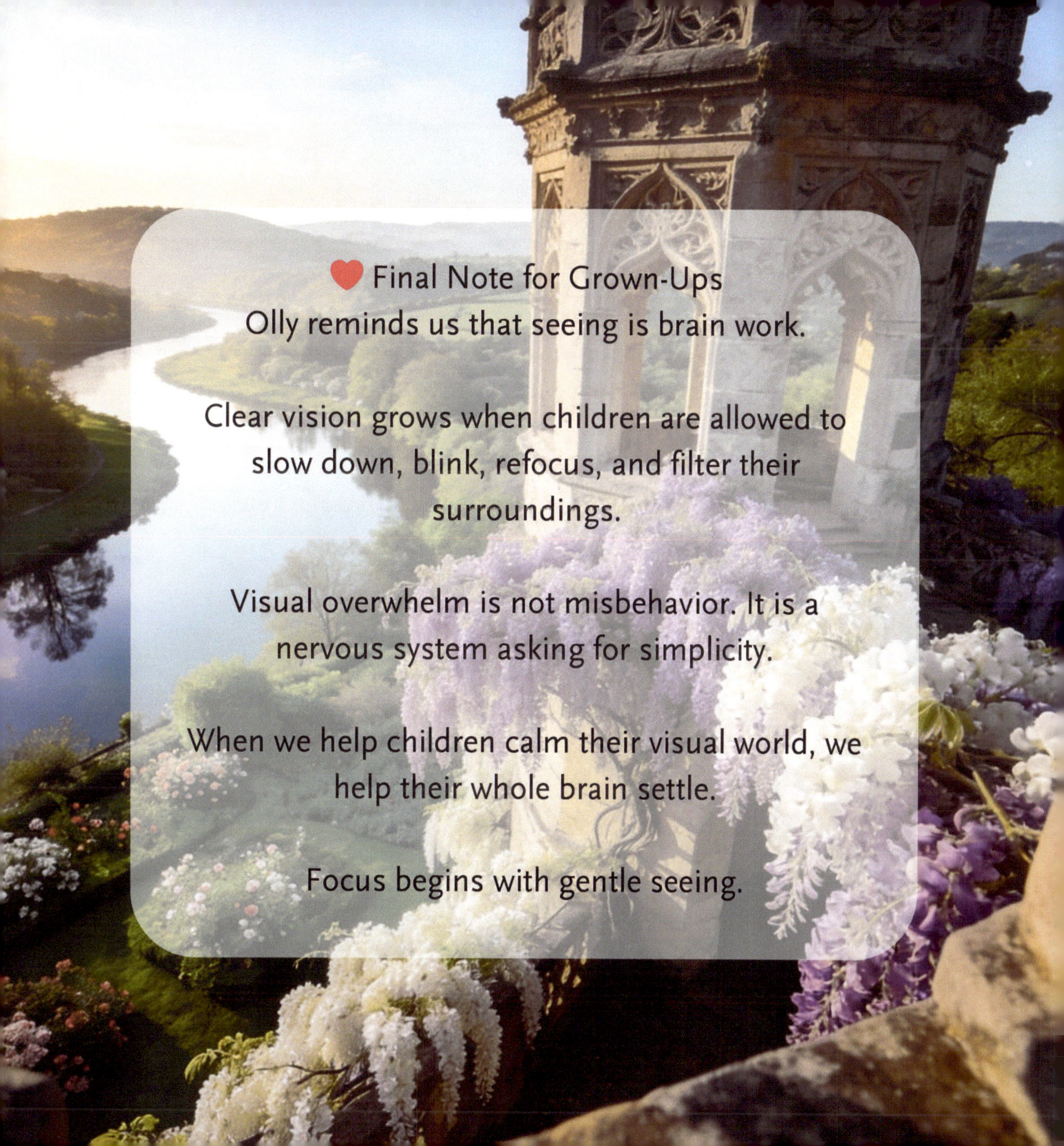

❤️ Final Note for Grown-Ups

Olly reminds us that seeing is brain work.

Clear vision grows when children are allowed to slow down, blink, refocus, and filter their surroundings.

Visual overwhelm is not misbehavior. It is a nervous system asking for simplicity.

When we help children calm their visual world, we help their whole brain settle.

Focus begins with gentle seeing.

Resources & GROWTH

Resources:

(Occipital lobe + vision, visual processing)

- National Institute of Neurological Disorders and Stroke (NINDS)
- MedlinePlus (U.S. National Library of Medicine)
- Encyclopedia Britannica (occipital lobe)
- NINDS (NIH). Brain Basics: Know Your Brain (occipital lobe, visual processing).
- Cleveland Clinic. Brain: parts and function (how the brain sees).
- NINDS (NIH). Brain Basics: Know Your Brain / Brain Basics resources. (occipital lobe/vision).
- Cleveland Clinic. Brain: parts and function (how the brain sees).
- CASEL. CASEL's SEL Framework (self-awareness, self-management, social awareness, relationship skills, responsible decision-making). (2020, updated).
- CDC. Child Development: Positive Parenting Tips and developmental guidance by age. (2024-2025).
- National Eye Institute (NIH). Research news and explainers referencing occipital cortex/visual processing. (2020-2023).

Growth:

(If you or someone you know wants to strengthen visual awareness or focus)

- American Academy of Ophthalmology (EyeSmart)
    - Look for: vision development
- Understood.org
    - Search: visual processing

Acknowledgments:

To my students: past, present, and future:
You taught me more than any textbook ever could. By watching how you explored, questioned, struggled, and discovered, I learned how minds truly grow. Every "aha" moment in these pages began with you.

To the parents, teachers, and guides walking beside children every day:
Thank you for listening, for advocating when things feel complicated, and for believing that understanding a child's inner world truly matters. Your patience and care shape futures in ways you may never fully see.

To my husband, my best friend, and my constant partner in learning:
Thank you for walking beside me, believing in this work, and holding space for creativity to exist. Sharing curiosity, growth, and creation with you has been one of the greatest honors of my life.

To the great thinkers who shaped how I see the world: including Dr. Seuss, Albert Einstein, Leonardo da Vinci, Stephen Hawking, Ludwig van Beethoven, just to name a few, and my English teacher, Mrs. Godfrey:
Your curiosity, courage, and imagination reminded me that learning is limitless, creativity is powerful, and questions matter. Mrs. Godfrey, your encouragement, belief in me, and love of words sparked my writing journey. If not for you, I may never have picked up a pen.

And to my publisher, NextGen Learners™:
It was by fate that we met, and by inspiration that we now move forward. Together, we build something meaningful. Together, we leave this legacy.

Lastly, to my own brain:
Complex, resilient, ever-adapting, and endlessly learning. Thank you for carrying stories, holding wonder, and reminding me that growth is always possible.

Maisel McLaula is the author of the Think Like a Brain!™ children's book collection, a story-first series that blends neuroscience, imagination, and emotional safety for young readers.

Drawing from decades of experience supporting children and families, her work centers on a simple but powerful belief: children deserve to understand and accept who they are, including how their brains work and why they make the choices they do.

Through playful characters and gentle metaphors, Maisel's stories help children discover the "why" behind their thoughts, feelings, and actions, turning confusion into clarity and self-judgment into self-compassion. By understanding their inner systems, children gain the power to respond with kindness toward themselves and to recognize those same patterns in others.

Maisel believes that when children feel safe inside their own minds, empathy grows naturally. The Think Like a Brain!™ collection offers a foundation for confidence, understanding, and care that grows with the child.

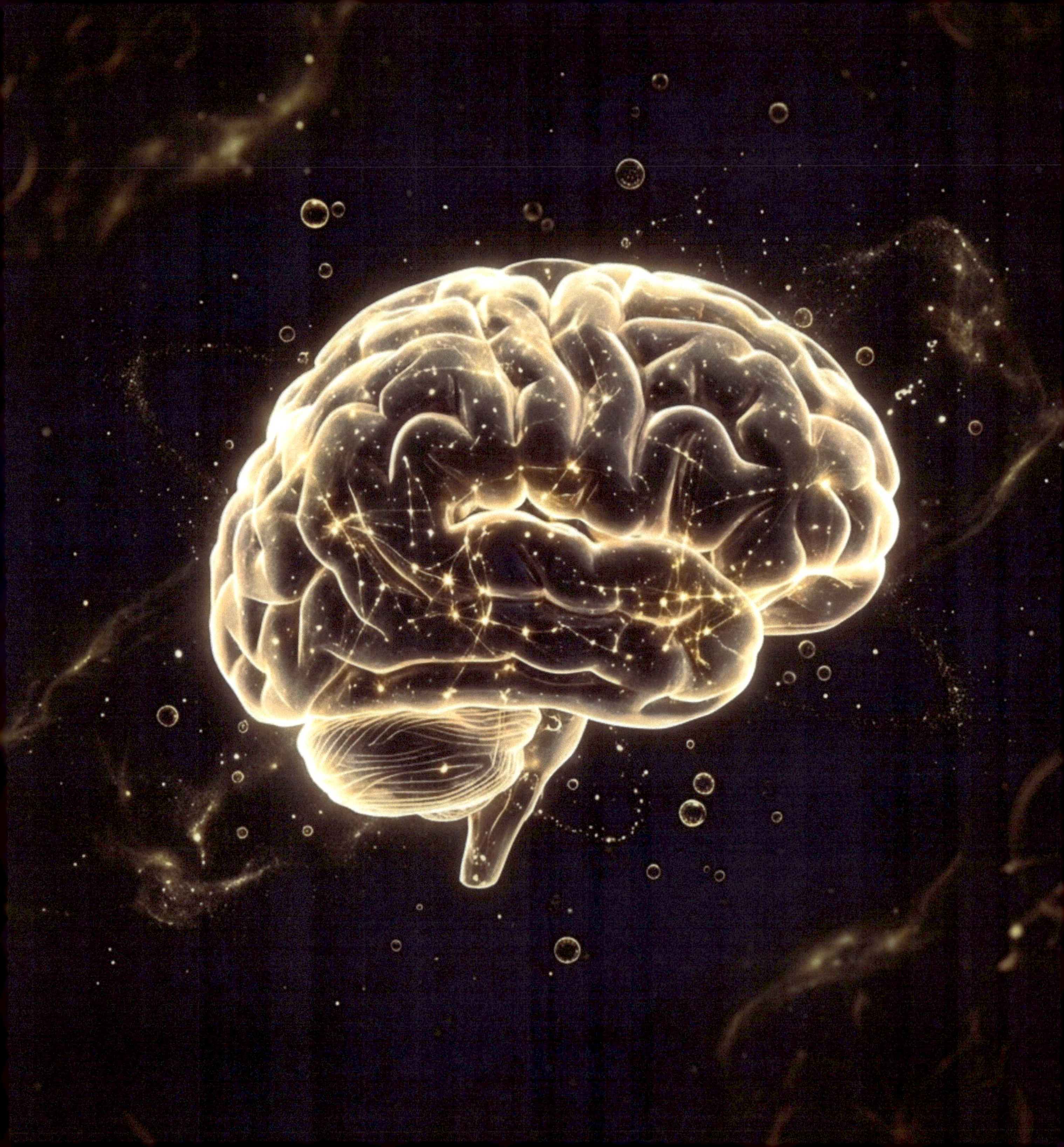

ABOUT THE PUBLISHER

NextGen Learners LLC is an educational publishing company dedicated to helping children understand how they think, feel, learn, and grow.

Through imaginative storytelling, neuroscience-informed concepts, and real-world application, NextGen Learners™ creates books and resources that support emotional development, critical thinking, self-awareness, and confidence in young minds.

Our work is designed not only for children, but for the parents, educators, caregivers, and professionals who walk beside them. Each book serves as a bridge, helping adults better understand the child in front of them, while helping children better understand themselves.

By creating shared language and meaningful connection points, our stories foster deeper relationships, reduce misunderstanding, and encourage acceptance, both within oneself and toward others.

We believe every child's mind is uniquely wired, and that understanding those differences is the foundation for compassion, growth, and lifelong learning.
With multiple collections designed to support cognitive, emotional, and developmental growth, NextGen Learners provides tools that children can carry with them, and tools that adults can use to guide them with clarity, patience, and purpose.

NG
NEXTGEN
LEARNERS
PRESS

MORE TO EXPLORE

The Brainy Bunch is just the beginning.
More stories, tools, and brain-friendly resources are on the way! Designed to help children, parents, and educators better understand how the brain works in everyday life.

FREE PRINTABLES

We've created printable pages to go along with this book, including:

-Reading Guides
-Brain helper visuals
-Character Cards

These are available for free as part of our growing resource library.

Our full website experience is built to include:

-Downloadable activity packs
-Parent & teacher guides
-Expanded learning tools
-Additional collections and topics

STAY CONNECTED

We love hearing how children connect with these stories!

If there is a topic, feeling, or challenge you would like to see explored in future books, we would love to hear from you!
Contact us: Inspire@NextGenLearners.com
NextGenLearners.com

Your ideas help shape future stories and support the growth of this collection.

This collection grows with you. New resources and stories are added regularly.

Maisel McLaula loves hearing from the tiny humans who read her books.

Did a story make you giggle, spark a big idea, or inspire you to draw your very own character? She would be delighted to see it! Send your letters and artwork to:

NextGen Learners
Attn: Maisel McLaula
8666 Beechmont Ave #1063
Cincinnati, OH 45255

Every letter is read with a smile, and your artwork just might make her whole day.

OPEN BRAIN

www.ingramcontent.com/pod-product-compliance
Lightning Source LLC
Chambersburg PA
CBHW041322210926
51847CB00002B/27
9781972860038